EPA 402/K-13/002 I September 2013 (revised)

Home Buyer's and Seller's Guide to Radon

Indoor Air Quality (IAQ)

EPA RECOMMENDS:

See page **20**

- ☐ If you are buying or selling a home, have it tested for radon.

- ☐ For a new home, ask if radon-resistant construction features were used and if the home has been tested.

- ☐ Fix the home if the radon level is 4 picocuries per liter (pCi/L) or higher.

- ☐ Radon levels less than 4 pCi/L still pose a risk and, in many cases, may be reduced.

- ☐ Take steps to prevent device interference when conducting a radon test.

EPA estimates that radon causes thousands of cancer deaths in the U.S. each year.

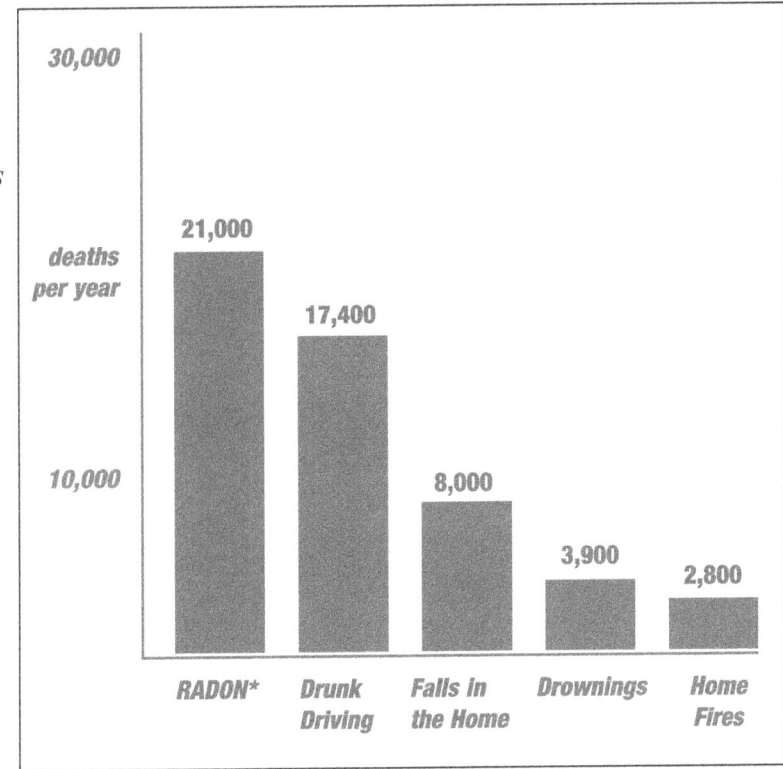

Radon is estimated to cause about 21,000 lung cancer deaths per year, according to EPA's 2003 Assessment of Risks from Radon in Homes (EPA 402-R-03-003). The numbers of deaths from other causes are taken from the Centers for Disease Control and Prevention's 2005-2006 National Center for Injury Prevention and Control Report and 2006 National Safety Council Reports.

EPA 402/K-13/002 I September 2013 (revised)

Table of Contents

EPA 402/K-13/002 | September 2013 (revised)

EPA 402/K-13/002 I September 2013 (revised)

Overview

This *Guide* answers important questions about radon and lung cancer risk. It also answers questions about testing and fixing for anyone buying or selling a home.

Radon Is a Cancer-Causing, Radioactive Gas

You cannot see, smell, or taste radon. But it still may be a problem in your home. When you breathe air containing radon, you increase your risk of getting lung cancer. In fact, the Surgeon General of the United States has warned that radon is the second leading cause of lung cancer in the United States today. *If you smoke and your home has high radon levels, your risk of lung cancer is especially high.*

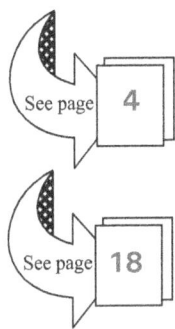

See page 4

See page 18

EPA Risk Assessment for Radon in Indoor Air

EPA has updated its estimate of the lung cancer risks from exposure to radon in indoor air. The Agency's updated risk assessment, *EPA Assessment of Risks from Radon in Homes* (EPA 402-R-03-003, June 2003), is available at http://www.epa.gov/radon/pdfs/402-r-03-003.pdf as a downloadable Adobe Acrobat PDF file. EPA's reassessment was based on the National Academy of Sciences' (NAS) report on the *Health Effects of Exposure to Radon* (BEIR VI, 1999). The Agency now estimates that there are about 21,000 annual radon-related lung cancer deaths, an estimate consistent with the NAS Report's findings.

You Should Test for Radon

Testing is the only way to find out your home's radon levels. EPA and the Surgeon General recommend testing all homes below the third floor for radon.

See page 20

You Can Fix a Radon Problem

If you find that you have high radon levels, there are ways to fix a radon problem. Even very high levels can be reduced to acceptable levels.

If You Are Selling a Home...

EPA recommends that you test your home before putting it on the market and, if necessary, lower your radon levels. Save the test results and all information you have about steps that were taken to fix any problems. This could be a positive selling point.

EPA 402/K-13/002 | September 2013 (revised)

If You Are Buying a Home...

EPA recommends that you know what the indoor radon level is in any home you consider buying. Ask the seller for their radon test results. If the home has a radon-reduction system, ask the seller for any information they have about the system.

If the home has not yet been tested, you should have the house tested.

See page **6**

If you are having a new home built, there are features that can be incorporated into your home during construction to reduce radon levels.

See page **9**

The radon testing guidelines in this *Guide* have been developed specifically to deal with the time-sensitive nature of home purchases and sales, and the potential for radon device interference. These guidelines are slightly different from the guidelines in other EPA publications which provide radon testing and reduction information for *non-real estate* situations.

This *Guide* recommends three short-term testing options for real estate transactions. EPA also recommends testing a home in the lowest level that could be used regularly, since a buyer may choose to live in a lower area of the home than that used by the seller.

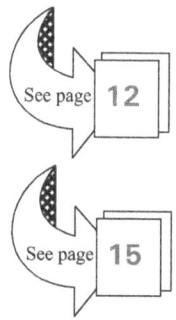
See page **12**

See page **15**

EPA 402/K-13/002 I September 2013 (revised)

1. Why Should I Test for Radon

a. Radon Has Been Found In Homes All Over the United States

Radon is a radioactive gas that has been found in homes all over the United States. It comes from the natural breakdown of uranium in soil, rock, and water and gets into the air you breathe. Radon typically moves up through the ground to the air above and into your home through cracks and other holes in the foundation. Radon can also enter your home through well water. Your home can trap radon inside.  See page 25

Any home can have a radon problem. This means new and old homes, well-sealed and drafty homes, and homes with or without basements. In fact, you and your family are most likely to get your greatest radiation exposure at home. That is where you spend most of your time.

Nearly one out of every 15 homes in the United States is estimated to have an elevated radon level (4 pCi/L or more). Elevated levels of radon gas have been found in homes in your state. Contact your state radon office for information about radon in your area. See page 29 34

EPA 402/K-13/002 I September 2013 (revised)

b. EPA and the Surgeon General Recommend That You Test Your Home

Testing is the only way to know if you and your family are at risk from radon. EPA and the Surgeon General recommend testing all homes below the third floor for radon.

U.S. SURGEON GENERAL HEALTH ADVISORY

"Indoor radon is the second-leading cause of lung cancer in the United States and breathing it over prolonged periods can present a significant health risk to families all over the country. It's important to know that this threat is completely preventable. Radon can be detected with a simple test and fixed through well-established venting techniques." January 2005

You cannot predict radon levels based on state, local, and neighborhood radon measurements. Do not rely on radon test results taken in other homes in the neighborhood to estimate the radon level in your home. Homes which are next to each other can have different indoor radon levels. Testing is the only way to find out what your home's radon level is.

In some areas, companies may offer different types of radon service agreements. Some agreements let you pay a one-time fee that covers both testing and radon mitigation, if needed. Contact your state radon office to find out if these are available in your state.

See page 34

EPA 402/K-13/002 I September 2013 (revised)

2. I'm Selling a Home. What Should I Do?

a. If Your Home Has Already Been Tested for Radon...

If you are thinking of selling your home and you have already tested your home for radon, review the *Radon Testing Checklist* to make sure that the test was done correctly. If so, provide your test results to the buyer.

 See page 20

No matter what kind of test was done, a potential buyer may ask for a new test, especially if:

☐ The *Radon Testing Checklist* items were not met;

☐ The last test is not recent, e.g., within two years;

☐ You have renovated or altered your home since you tested; or

☐ The buyer plans to use a lower level of the house than was tested, such as a basement that could be used regularly by the buyer.

A buyer may also ask for a new test if your state or local government requires disclosure of radon information to buyers.

EPA 402/K-13/002 I September 2013 (revised)

b. If Your Home Has *Not Yet* Been Tested for Radon...

Have a test taken as soon as possible. If you can, test your home before putting it on the market. You should test in the lowest level of the home that could be used regularly. This means testing in the lowest level that you currently live in or a lower level not currently used, but which a buyer might use as a family room or play area, etc.

The radon test result is important information about your home's radon level. Some states require radon measurement testers to follow a specific testing protocol. If you do the test yourself, you should carefully follow the testing protocol for your area or EPA's *Radon Testing Checklist*. If you hire a contractor to test your residence, protect yourself by hiring a **qualified***
individual or company.

See page 20 34

You can determine a service provider's qualifications to perform radon measurements or to mitigate your home in several ways. **Check with your state radon office**. Many states require radon professionals to be licensed, certified, or registered. Most states can provide you with a list of knowledgeable radon service providers doing business in the state. In states that don't regulate radon services, **ask the contractor if they hold a professional proficiency or certification credential**. Such programs usually provide members with a photo-ID card, which indicates their qualification(s) and its expiration date. If in doubt, you should check with their credentialing organization. Alternatively, **ask the contractor if they've successfully completed formal training** appropriate for testing or mitigation, e.g., a course in radon measurement or radon mitigation.

* You should first call your state radon office for information on qualified radon service providers and state-specific radon measurement or mitigation requirements. For up-to-date information on state radon program offices, visit **http://www.epa.gov/radon/whereyoulive.html**. EPA's detailed and technical guidance on radon measurement and mitigation is included in Section 8 (p. 29); however, state requirements or guidance may be more stringent. Visit **http://www.epa.gov/radon/radontest.html** for links to private sector radon credentialing programs.

EPA 402/K-13/002 | September 2013 (revised)

3. I'm Buying a Home. What Should I Do?

a. If the Home Has Already Been Tested for Radon...

If you are thinking of buying a home, you may decide to accept an earlier test result from the seller or ask the seller for a new test to be conducted by a qualified radon tester. Before you accept the seller's test, you should determine:

☐ The results of previous testing;

☐ Who conducted the previous test: the homeowner, a radon professional, or some other person;

☐ Where in the home the previous test was taken, especially if you may plan to live in a lower level of the home. For example, the test may have been taken on the first floor. However, if you want to use the basement as living space, test there; and

☐ What, if any, structural changes, alterations, or changes in the heating, ventilation, and air conditioning (HVAC) system have been made to the house since the test was done. Such changes might affect radon levels.

If you accept the seller's test, make sure that the test followed the *Radon Testing Checklist*.

If you decide that a new test is needed, discuss it with the seller as soon as possible. If you decide to use a qualified radon tester, contact your state radon office to obtain a copy of their approved list of radon testing companies.

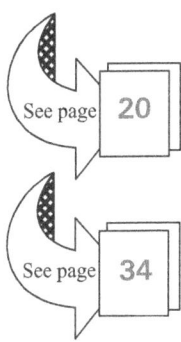

See page 20

See page 34

EPA 402/K-13/002 | September 2013 (revised)

b. If the Home Has *Not Yet* Been Tested for Radon...

Make sure that a radon test is done as soon as possible. Consider including provisions in the contract specifying:

☐ Where the test will be located;

☐ Who should conduct the test;

See page **12**

☐ What type of test to do;

☐ When to do the test;

See page **14**

☐ How the seller and the buyer will share the test results and test costs (if necessary); and

☐ When radon mitigation measures will be taken, and who will pay for them.

Make sure that the test is done in the lowest level of the home that could be used regularly. This means the lowest level that you are going to use as living space whether it is finished or unfinished. A state or local radon official or qualified radon tester can help you make some of these decisions.

If you decide to finish or renovate an unfinished area of the home in the future, a radon test should be done before starting the project and after the project is finished. Generally, it is less expensive to install a radon-reduction system before (or during) renovations rather than afterwards.

EPA 402/K-13/002 I September 2013 (revised)

4. I'm Buying or Building a New Home. How Can I Protect My Family?

a. Why Should I Buy a Radon-Resistant Home?

Radon-resistant techniques work. When installed properly and completely, these simple and inexpensive passive techniques can help to reduce radon levels. In addition, installing them at the time of construction makes it easier to reduce radon levels further if the passive techniques don't reduce radon levels to below 4 pCi/L. Radon-resistant techniques may also help to lower moisture levels and those of other soil gases. Radon-resistant techniques:

✔ *Make Upgrading Easy*: Even if built to be radon-resistant, **every new home should be tested for radon as soon as possible after occupancy**. If you have a test result of 4 pCi/L or more, a vent fan can easily be added to the passive system to make it an active system and further reduce radon levels.

✔ *Are Cost-Effective*: Building radon-resistant features into the house during construction is easier and cheaper than fixing a radon problem from scratch later. Let your builder know that radon-resistant features are easy to install using common building materials.

✔ *Save Money*: When installed properly and completely, radon-resistant techniques can also make your home more energy efficient and help you save on your energy costs.

Including passive radon-resistant features in a **new home** during construction usually costs less than fixing the home later. If your radon level is 4 pCi/L or more, consult a qualified mitigator to estimate the cost of upgrading to an active system by adding a vent fan to reduce the radon level. In an **existing home,** the cost to install a radon mitigation system is about the same as for other common home repairs. Check with, and get an estimate from, one or more qualified mitigators before fixing.

EPA 402/K-13/002 | September 2013 (revised)

b. What Are Radon-Resistant Features?

Radon-resistant techniques (features) may vary for different foundations and site requirements. If you're having a house built, ask your builder if they're using a recognized approach (International Residential Code, Appendix F, ASTM E 1465-08, and ANSI/AARST RRNC 2.0 as examples). If your new house was built (or will be built) to be radon-resistant, it will include these basic elements:

See page 32

1. **Gas-Permeable Layer:** This layer is placed beneath the slab or flooring system to allow the soil gas to move freely underneath the house. In many cases, the material used is a 4-inch layer of clean gravel. This gas-permeable layer is used only in homes with basement and slab-on-grade foundations; it is not used in homes with crawlspace foundations.

2. **Plastic Sheeting:** Plastic sheeting is placed on top of the gas-permeable layer and under the slab to help prevent the soil gas from entering the home. In crawl spaces, the sheeting (with seams sealed) is placed directly over the crawlspace floor.

3. **Sealing and Caulking:** All below-grade openings in the foundation and walls are sealed to reduce soil gas entry into the home.

4. **Vent Pipe:** A 3- or 4-inch PVC pipe (or other gas-tight pipe) runs from the gas-permeable layer through the house to the roof, to safely vent radon and other soil gases to the outside.

5. **Junction Boxes:** An electrical junction box is included in the attic to make the wiring and installation of a vent fan easier. For example, you decide to activate the passive system because your test result showed an elevated radon level (4 pCi/L or more). A separate junction box is placed in the living space to power the vent fan alarm. An alarm is installed along with the vent fan to indicate when the vent fan is not operating properly.

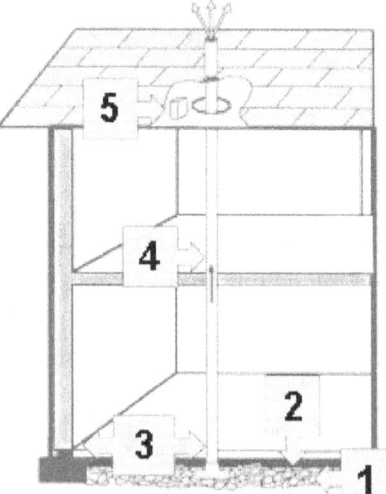

EPA 402/K-13/002 I September 2013 (revised)

5. How Can I Get Reliable Radon Test Results?

Radon testing is easy and the only way to find out if you have a radon problem in your home.

a. Types of Radon Devices

Since you cannot see or smell radon, special equipment is needed to detect it. When you're ready to test your home, you can order a radon test kit by mail from a qualified radon measurement services provider or laboratory. You can also hire a qualified radon tester, very often a home inspector, who will use a radon device(s) suitable to your situation. The most common types of radon testing devices are listed below. As new testing devices are developed, you may want to check with your state radon office before you test to get the most up-to-date information.

See page 16 30

See page 34

✔ Passive Devices

Passive radon testing devices do not need power to function. These include **charcoal canisters, alpha-track detectors, charcoal liquid scintillation devices,** and **electret ion chamber detectors,** which are available in hardware, drug, and other stores; they can also be ordered by mail or phone. These devices are exposed to the air in the home for a specified period of time and then sent to a laboratory for analysis. Both short-term and long-term passive devices are generally inexpensive. Some of these devices may have features that offer more resistance to test interference or disturbance than other passive devices. Qualified radon testers may use any of these devices to measure the home's radon level.

EPA 402/K-13/002 I September 2013 (revised)

✔ *Active Devices*

Active radon testing devices require power to function. These include **continuous radon monitors** and **continuous working level monitors**. They continuously measure and record the amount of radon or its decay products in the air. Many of these devices provide a report of this information which can reveal any unusual or abnormal swings in the radon level during the test period. A qualified tester can explain this report to you. In addition, some of these devices are specifically designed to deter and detect test interference. Some technically advanced active devices offer anti-interference features. Although these tests may cost more, they may ensure a more reliable result.

b. General Information for All Devices

A state or local radon official can explain the differences between devices and recommend the ones which are most appropriate for your needs and expected testing conditions.

Make sure to use a radon measurement device from a qualified laboratory. Certain precautions should be followed to avoid interference during the test period; see the *Radon Testing Checklist* for more information on how to get a reliable test result.

See page 20

Radon Test Device Placement

EPA recommends that the test device(s) be placed in the lowest level of the home that could be used regularly, whether it is finished or unfinished. Conduct the test in any space that could be used by the buyer as a bedroom, play area, family room, den, exercise room, or workshop. Based on their client's intended use of the space, the qualified testing professional should identify the appropriate test location and inform their client (buyer). Do **not** test in a closet, stairway, hallway, crawl space or in an enclosed area of high humidity or high air velocity. An enclosed area may include a kitchen, bathroom, laundry room or furnace room.

EPA 402/K-13/002 I September 2013 (revised)

c. Preventing or Detecting Test Interference

There is a potential for test interference in real estate transactions. There are several ways to prevent or detect test interference:

☐ Use a test device that frequently records radon or decay product levels to detect unusual swings;

☐ Employ a motion detector to determine whether the test device has been moved or if testing conditions have changed;

☐ Use a proximity detector to reveal the presence of people in the room which may correlate to possible changes in radon levels during the test;

☐ Record the barometric pressure to identify weather conditions which may have affected the test;

☐ Record the temperature to help assess whether doors and windows have been opened;

☐ Apply tamper-proof seals to windows to ensure closed-house conditions; and

See page 20

☐ Have the seller/occupant sign a non-interference agreement.

Home buyers and sellers should consult a qualified radon test provider about the use of these precautions.

EPA 402/K-13/002 I September 2013 (revised)

d. Length of Time to Test

Because radon levels tend to vary from day to day and season to season, a short-term test is less likely than a long-term test to tell you your year-round average radon level. However, if you need results quickly, a short-term test may be used to decide whether to fix the home.

There Are Two General Ways to Test Your Home for Radon:

✔ *Short-Term Testing*

The quickest way to test is with short-term tests. Short-term tests remain in your home from two to 90 days, depending on the device. There are two groups of devices which are more commonly used for short-term testing. The passive device group includes **alpha track detectors, charcoal canisters, charcoal liquid scintillation detectors,** and **electret ion chambers**. The active device group consists of different types of **continuous monitors**.

Whether you test for radon yourself or hire a qualified tester, all radon tests should be taken for a minimum of 48 hours. Some devices require a longer (minimum) length of time, e.g., a 7-day charcoal canister device.

✔ *Long-Term Testing*

Long-term tests remain in your home for more than 90 days. **Alpha track** and **electret ion chamber detectors** are commonly used for this type of testing. A long-term test result is more likely to tell you your home's year-round average radon level than a short-term test. If time permits (more than 90 days), long-term tests can be used to confirm initial short-term results. When long-term test results are 4 pCi/L or higher, EPA recommends fixing the home.

EPA 402/K-13/002 I September 2013 (revised)

e. Doing a Short-Term Test...

If you are testing in a real estate transaction and you need results quickly, any of the following three options for short-term tests are acceptable in determining whether the home should be fixed. Any real estate test for radon should include steps to prevent or detect interference with the test device.

See page 13

When Choosing a Short-Term Testing Option...

There are trade-offs among the short-term testing options. Two tests taken at the same time (simultaneous) would improve the precision of this radon test. One test followed by another test (sequential) would most likely give a better representation of the seasonal average. Both active and passive devices may have features which help to prevent test interference. Your state radon office can help you decide which option is best.

Short-Term Testing Options

What to Do Next

Passive:
Take two short-term tests at the same time in the same location for at least 48 hours.

Fix the home if the average of the two tests is 4 pCi/L or more.

or

Take an initial short-term test for at least 48 hours. Immediately upon completing the first test, do a second test using an identical device in the same location as the first test.

Fix the home if the average of the two tests is 4 pCi/L or more.

Active:
Test the home with a continuous monitor for at least 48 hours.

Fix the home if the average radon level is 4 pCi/L or more.

EPA 402/K-13/002 I September 2013 (revised)

f. Using Testing Devices Properly for Reliable Results

✔ *If You Do the Test Yourself*

When you are taking a short-term test, close windows and doors to the outside and keep them closed, except for normal entry and exit. If you are taking a short-term test lasting less than four days, be sure to:

☐ Close your windows and outside doors at least 12 hours before beginning the test;

☐ Do not conduct short-term tests lasting less than four days during severe storms or periods of high winds;

☐ Follow the testing instructions and record the start time and date;

☐ Place the test device at least 20 inches above the floor in a location where it will not be disturbed and where it will be away from drafts, high heat, high humidity, and exterior walls;

☐ Leave the test kit in place for as long as the test instructions say; and

☐ Once the test is finished, record the stop time and date, reseal the package, and return it immediately to the lab specified on the package for analysis.

You should receive your test results within a few days or weeks. If you need results quickly, you should find out how long results will take and, if necessary, request expedited service.

EPA 402/K-13/002 | September 2013 (revised)

✔ *If You Hire a Qualified Radon Tester*

In many cases, home buyers and sellers may decide to have the radon test done by a qualified radon tester who knows the proper conditions, test devices, and guidelines for obtaining a reliable radon test result. They can also:

See page 34

☐ Evaluate the home and recommend a testing approach designed to make sure you get reliable results;

☐ Explain how proper conditions can be maintained during the radon test;

☐ Emphasize to a home's occupants that a reliable test result depends upon their cooperation. Interference with, or disturbance of, the test or closed-house conditions will invalidate the test result;

☐ Analyze the data and report the measurement results; and

☐ Provide an independent test result.

Your state radon office may also have information about qualified radon testers and certification requirements.

g. Interpreting Radon Test Results

The average indoor radon level is estimated to be about 1.3 pCi/L; roughly 0.4 pCi/L of radon is normally found in the outside air. The U.S. Congress has set a long-term goal that indoor radon levels be no more than outdoor levels. While this goal is not yet technologically achievable for all homes, radon levels in many homes *can* be reduced to 2 pCi/L or less. A radon level below 4 pCi/L still poses a risk. Consider fixing when the radon level is between 2 and 4 pCi/L.

EPA 402/K-13/002 I September 2013 (revised)

Radon and Smoking

RADON RISK IF YOU SMOKE

Radon Level	If 1,000 people who smoked were exposed to this level over a lifetime*. . .	The risk of cancer from radon exposure compares to**. . .	WHAT TO DO: Stop Smoking and. . .
20 pCi/L	About 260 people could get lung cancer	◄ 250 times the risk of drowning	Fix your home
10 pCi/L	About 150 people could get lung cancer	◄ 200 times the risk of dying in a home fire	Fix your home
8 pCi/L	About 120 people could get lung cancer	◄ 30 times the risk of dying in a fall	Fix your home
4 pCi/L	About 62 people could get lung cancer	◄ 5 times the risk of dying in a car crash	Fix your home
2 pCi/L	About 32 people could get lung cancer	◄ 6 times the risk of dying from poison	Consider fixing between 2 and 4 pCi/L
1.3 pCi/L	About 20 people could get lung cancer	(Average indoor radon level)	(Reducing radon levels below
0.4 pCi/L	About 3 people could get lung cancer	(Average outdoor radon level)	2 pCi/L is difficult)

Note: If you are a former smoker, your risk may be lower.

RADON RISK IF YOU HAVE NEVER SMOKED

Radon Level	If 1,000 people who never smoked were exposed to this level over a lifetime*. . .	The risk of cancer from radon exposure compares to**. . .	WHAT TO DO:
20 pCi/L	About 36 people could get lung cancer	◄ 35 times the risk of drowning	Fix your home
10 pCi/L	About 18 people could get lung cancer	◄ 20 times the risk of dying in a home fire	Fix your home
8 pCi/L	About 15 people could get lung cancer	◄ 4 times the risk of dying in a fall	Fix your home
4 pCi/L	About 7 people could get lung cancer	◄ The risk of dying in a car crash	Fix your home
2 pCi/L	About 4 people could get lung cancer	◄ The risk of dying from poison	Consider fixing between 2 and 4 pCi/L
1.3 pCi/L	About 2 people could get lung cancer	(Average indoor radon level)	(Reducing radon levels below
0.4 pCi/L		(Average outdoor radon level)	2 pCi/L is difficult)

Note: If you are a former smoker, your risk may be higher.
*Lifetime risk of lung cancer deaths from *EPA Assessment of Risks from Radon in Homes* (EPA 402-R-03-003).
**Comparison data calculated using the Centers for Disease Control and Prevention's 1999-2001 National Center for Injury Prevention and Control Reports.

EPA 402/K-13/002 | September 2013 (revised)

Sometimes short-term tests are less definitive about whether the radon level in the home is at or above 4 pCi/L; particularly when the results are close to 4 pCi/L. For example, if the average of two short-term tests is 4.1 pCi/L, there is about a 50 percent chance that the year-round average is somewhat below, or above, 4 pCi/L.

However, EPA believes that any radon exposure carries some risk; no level of radon is safe. Even radon levels below 4 pCi/L pose some risk. You can reduce your risk of lung cancer by lowering your radon level.

As with other environmental pollutants, there is some uncertainty about the magnitude of radon health risks. However, we know more about radon risks than risks from most other cancer-causing substances. This is because estimates of radon risks are based on data from human studies (underground miners). Additional studies on more typical populations are under way.

Your radon measurement will give you an idea of your risk of getting lung cancer. Your chances of getting lung cancer from radon depend mostly on:

✔ Your home's radon level;

✔ The amount of time you spend in your home; and

✔ Whether you are a smoker or have ever smoked.

Smoking combined with radon is an especially serious health risk. If you smoke or are a former smoker, the presence of radon greatly increases your risk of lung cancer. If you stop smoking now and lower the radon level in your house, you will reduce your lung cancer risk.

Radon Testing Checklist

For reliable test results, follow this *Radon Testing Checklist* carefully. Testing for radon is not complicated. Improper testing may yield inaccurate results and require another test. Disturbing or interfering with the test device, or with **closed-house conditions***, may invalidate the test results and is illegal in some states. If the seller or qualified tester cannot confirm that all items have been completed, take another test.

✔ *Before Conducting a Radon Test*:

☐ Notify the occupants of the importance of proper testing conditions. Give the occupants written instructions or a copy of this *Guide* and explain the directions carefully.

☐ Conduct the radon test for a minimum of 48 hours; some test devices have a minimum exposure time greater than 48 hours.

☐ When doing a short-term test ranging from 2-4 days, it is important to maintain closed-house conditions for at least 12 hours before the beginning of the test and during the entire test period.

☐ When doing a short-term test ranging from 4-7 days, EPA recommends that closed-house conditions be maintained.

☐ If you conduct the test yourself, use a qualified radon measurement device and follow the laboratory's instructions. Your state may be able to provide you with a list of do-it-yourself test devices available from qualified laboratories.

☐ If you hire someone to do the test, hire only a qualified individual. Some states issue photo identification (ID) cards; ask to see it. The tester's ID number, if available, should be included or noted in the test report.

*Closed-house conditions means keeping all windows closed, keeping doors closed except for normal entry and exit, and not operating fans or other machines which bring in air from outside. Fans that are part of a radon-reduction system or small exhaust fans operating for only short periods of time may run during the test.

EPA 402/K-13/002 I September 2013 (revised)

Radon Testing Checklist

(continued)

☐ The test should include method(s) to prevent or detect interference with testing conditions or with the testing device itself.

☐ If the house has an active radon-reduction system, make sure the vent fan is operating properly. If the fan is not operating properly, have it (or ask to have it) repaired and then test.

✔ *During a Radon Test*:

☐ Maintain closed-house conditions during the entire duration of a short-term test, especially for tests shorter than one week in length.

☐ Operate the home's heating and cooling systems normally during the test. For tests lasting less than one week, operate only air-conditioning units which recirculate interior air.

☐ Do not disturb the test device at any time during the test.

☐ If a radon-reduction system is in place, make sure the system is working properly and will be in operation during the entire radon test.

✔ *After a Radon Test*:

☐ If you conduct the test yourself, be sure to promptly return the test device to the laboratory. Be sure to complete the required information, including start and stop times, test location, etc.

☐ If an elevated radon level is found, fix the home. Contact a qualified radon-reduction contractor about lowering the radon level. EPA recommends that you fix the home when the radon level is 4 pCi/L or more.

☐ Be sure that you or the radon tester can demonstrate or provide information to ensure that the testing conditions were not violated during the testing period.

6. What Should I Do if the Radon Level is High?

a. High Radon Levels Can Be Reduced

EPA recommends that you take action to reduce your home's indoor radon levels if your radon test result is 4 pCi/L or higher. It is better to correct a radon problem before placing your home on the market because then you have more time to address a radon problem.

If elevated levels are found during the real estate transaction, the buyer and seller should discuss the timing and costs of radon reduction. The cost of making repairs to reduce radon levels depends on how your home was built and other factors. Most homes can be fixed for about the same cost as other common home repairs. Check with and get an estimate from one or more qualified mitigators.

b. How to Lower the Radon Level In Your Home

A variety of methods can be used to reduce radon in homes. Sealing cracks and other openings in the foundation is a basic part of most approaches to radon reduction. EPA does not recommend the use of sealing alone to limit radon entry. Sealing alone has not been shown to lower radon levels significantly or consistently.

In most cases, a system with a vent pipe(s) and fan(s) is used to reduce radon. These "sub-slab depressurization" systems do not require major changes to your home. Similar systems can also be installed in homes with crawl spaces. These systems prevent radon gas from entering the home from below the concrete floor and from outside the foundation. Radon mitigation contractors may use other methods that may also work in your home. The right system depends on the design of your home and other factors.

EPA 402/K-13/002 I September 2013 (revised)

Techniques for reducing radon are discussed in EPA's *Consumer's Guide to Radon Reduction*. As with any other household appliance, there are costs associated with the operation of a radon-reduction system.

See page 32

Radon and Home Renovations

If you are planning any major renovations, such as converting an unfinished basement area into living space, it is especially important to test the area before you begin.

If your test results indicate an elevated radon level, radon-resistant techniques can be inexpensively included as part of the renovation. Major renovations can change the level of radon in any home. Test again after the work is completed.

You should also test your home again after it is fixed to be sure that radon levels have been reduced. If your living patterns change and you begin occupying a lower level of your home (such as a basement), you should retest your home on that level. In addition, it is a good idea to retest your home sometime in the future to be sure radon levels remain low.

EPA 402/K-13/002 | September 2013 (revised)

c. Selecting a Radon-Reduction (Mitigation) Contractor

Select a qualified radon-reduction contractor to reduce the radon level in your home. Any mitigation measures taken or system installed in your home must conform to your state's regulations. In states without regulations covering mitigation, EPA recommends that the system conform to ASTM E 2121.

See page 32 33

EPA recommends that the mitigation contractor review the radon measurement results before beginning any radon-reduction work. Test again after the radon mitigation work has been completed to confirm that previous elevated levels have been reduced. EPA recommends that the test be conducted by an independent, qualified radon tester.

d. What Can a Qualified Radon-Reduction Contractor Do for You?

A qualified radon-reduction (mitigation) contractor should be able to:

- ☐ Review testing guidelines and measurement results, and determine if additional measurements are needed;

- ☐ Evaluate the radon problem and provide you with a detailed, written proposal on how radon levels will be lowered;

- ☐ Design a radon-reduction system;

- ☐ Install the system according to EPA recommended standard, or state and local codes; and

- ☐ Make sure the finished system effectively reduces radon levels to acceptable levels.

EPA 402/K-13/002 | September 2013 (revised)

Choose a radon mitigation contractor to fix your radon problem just as you would for any other home repair. You may want to get more than one estimate, and ask for and check their references. Make sure the person you hire is qualified to install a mitigation system. Some states regulate or certify radon mitigation services providers.

Be aware that a potential conflict of interest exists if the same person or firm performs the testing and installs the mitigation system. Some states may require the homeowner to sign a waiver in such cases. If the same person or firm does the testing and mitigation, make sure the testing is done in accordance with the *Radon Testing Checklist*. Contact your state radon office for more information.

e. Radon In Water

The radon in your home's indoor air can come from two sources, the soil or your water supply. Compared to radon entering your home through the water, radon entering your home through the soil is a much larger risk. If you've tested for radon in air and have elevated radon levels **and** your water comes from a private well, have your water tested. The devices and procedures for testing your home's water supply are different from those used for measuring radon in air.

The radon in your water supply poses an inhalation risk and an ingestion risk. Research has shown that your risk of lung cancer from breathing radon in air is much larger than your risk of stomach cancer from swallowing water with radon in it. Most of your risk from radon in water comes from radon released into the air when water is used for showering and other household purposes.

Radon in your home's water is not usually a problem when its source is surface water. A radon in water problem is more likely when its source is ground water, e.g., a private well or a public water supply system that uses ground water. Some public water systems treat their water to reduce radon levels before it is delivered to your home. If you are concerned that radon may be entering your home through the water and your water comes from a public water supply, contact your water supplier.

EPA 402/K-13/002 | September 2013 (revised)

If you've tested your private well and have a radon in water problem, it can be fixed. Your home's water supply can be treated in one of two ways. **Point-of-entry** treatment can effectively remove radon from the water before it enters your home. Point-of-entry treatment usually employs either granular activated carbon (GAC) filters or aeration devices. While GAC filters usually cost less than aeration devices, filters can collect radioactivity and may require a special method of disposal. **Point-of-use** treatment devices remove radon from your water at the tap, but only treat a small portion of the water you use, e.g., the water you drink. Point-of-use devices are not effective in reducing the risk from breathing radon released into the air from all water used in the home.

For information on radon in water, testing and treatment, and existing or planned radon in drinking water standards, or for general help, call EPA's Drinking Water Hotline at (800) 426-4791 or visit **http://water.epa.gov/lawsregs/rulesregs/sdwa/ radon/index.cfm**, an EPA web site. If your water comes from a private well, you can also contact your state radon office.

EPA 402/K-13/002 I September 2013 (revised)

7. Radon Myths and Facts

MYTH #1: Scientists are not sure that radon really is a problem.

FACT: Although some scientists dispute the precise number of deaths due to radon, all the major health organizations (like the Centers for Disease Control, the American Lung Association, and the American Medical Association) agree with estimates that radon causes thousands of preventable lung cancer deaths every year. This is especially true among smokers, since the risk to smokers is much greater than to non-smokers.

See page 18

MYTH #2: Radon testing devices are not reliable and are difficult to find.

FACT: Reliable radon tests are available from qualified radon testers and companies. Active radon devices can continuously gather and periodically record radon levels to reveal any unusual swings in the radon level during the test. Reliable testing devices are also available by phone or mail-order, and can be purchased in hardware stores and other retail outlets. Contact your state radon office for a list of qualified radon test companies.

See page 34

MYTH #3: Radon testing is difficult and time-consuming.

FACT: Radon testing is easy. You can test your home yourself or hire a qualified radon test company. Either approach takes only a small amount of time and effort.

See page 6

MYTH #4: Homes with radon problems cannot be fixed.

FACT: There are solutions to radon problems in homes. Thousands of home owners have already lowered their radon levels. Most homes can be fixed for about the same cost as other common home repairs. Contact your state radon office for a list of qualified mitigation contractors.

MYTH #5: Radon only affects certain types of homes.

FACT: Radon can be a problem in all types of homes, including old homes, new homes, drafty homes, insulated homes, homes with basements, and homes without basements. Local geology, construction materials, and how the home was built are among the factors that can affect radon levels in homes.

EPA 402/K-13/002 I September 2013 (revised)

MYTH #6: Radon is only a problem in certain parts of the country.

FACT: High radon levels have been found in every state. Radon problems do vary from area to area, but the only way to know a home's radon level is to test.

MYTH #7: A neighbor's test result is a good indication of whether your home has a radon problem.

FACT: It is not. Radon levels vary from home to home. The only way to know if your home has a radon problem is to test it.

MYTH #8: Everyone should test their water for radon.

FACT: While radon gets into some homes through the water, it is important to first test the air in the home for radon. If your water comes from a public water system that uses ground water, call your water supplier. If high radon levels are found and the home has a private well, call the Safe Drinking Water Hotline at (800) 426-4791 for information on testing your water. Also, call your state radon office for more information about radon in air.

See page 25

MYTH #9: It is difficult to sell a home where radon problems have been discovered.

FACT: Where radon problems have been fixed, home sales have not been blocked. The added protection will be a good selling point.

MYTH #10: I have lived in my home for so long, it does not make sense to take action now.

FACT: You will reduce your risk of lung cancer when you reduce radon levels, even if you have lived with an elevated radon level for a long time.

MYTH #11: Short-term tests cannot be used for making a decision about whether to reduce the home's high radon levels.

FACT: Short-term tests can be used to decide whether to reduce the home's high radon levels. However, the closer the short-term testing result is to 4 pCi/L, the less certainty there is about whether the home's year-round average is above or below that level. Keep in mind that radon levels below 4 pCi/L still pose some risk and that radon levels can be reduced to 2 pCi/L or below in most homes.

EPA 402/K-13/002 I September 2013 (revised)

8. Need More Information about Radon?

If you have a radon-related question, you should contact your state radon office. The following web sites, hotlines, and publications are your best sources of information. Visit our Frequent Questions web site at **http://iaq.supportportal.com**. You can also find indoor air quality information and publications on EPA's many web sites.

a. World Wide Web Sites (EPA)

These are EPA's most important web sites for information on radon and indoor air quality in homes. All the EPA publications listed in this section are available on EPA's web sites.

- ☐ **http://www.epa.gov/radon/**. EPA's main radon page. Includes links to the NAS radon report, radon-resistant new construction, the map of radon zones, radon publications, hotlines, and more.

- ☐ **http://www.epa.gov/iaq/whereyoulive.html**. Provides detailed information on contacting your state's radon office, including links to some state web sites. State indoor air quality contacts are also included.

- ☐ **http://www.epa.gov/iaq/radon/pubs/index.html**. Offers the full text version of EPA's most popular radon publications, including the *Home Buyer's and Seller's Guide to Radon*, the *Consumer's Guide to Radon Reduction*, and the *Model Standards and Techniques for Control of Radon in New Residential Buildings*, and others.

- ☐ **http://www.epa.gov/iaq**. EPA's main page on indoor air quality. Includes information on indoor risk factors, e.g., asthma, secondhand smoke, carbon monoxide, duct cleaning, ozone generating devices, indoor air cleaners, flood cleanup, etc.

- ☐ **http://water.epa.gov/lawsregs/rulesregs/sdwa/radon/index.cfm**. EPA's main page on radon in water. Includes information on statutory requirements and links to the drinking water standards program.

b. Radon Hotlines (Toll-Free)

EPA supports the following hotlines to best serve consumers with radon-related questions and concerns.

☎ **1-800-SOS-RADON (767-7236).*** Purchase radon test kits by phone.

☎ **1-800-55RADON (557-2366).*** Get live help for your radon questions.

☎ **1-800-644-6999.*** Radon Fix-It Hotline. For general information on fixing or reducing the radon level in your home.

☎ **1-866-528-3187.*** Línea Directa de Información sobre Radón en Español. Hay operadores disponibles desde las 9:00 AM hasta las 5:00 PM para darle información sobre radón y como ordenar un kit para hacer la prueba de radón en su hogar.

☎ **1-800-426-4791.** Safe Drinking Water Hotline. For general information on drinking water, radon in water, testing and treatment, and standards for radon in drinking water. Operated under a contract with EPA.

*Operated by Kansas State University in partnership with EPA.

c. Printed Documents

Radon Risk and Testing

☐ *Home Buyer's and Seller's Guide to Radon*
(EPA 402/K-09/002, January 2009).
Everything you need to know about effectively dealing with radon during a residential real estate transaction. This publication can be viewed at **http://www.epa.gov/radon/pubs/hmbyguid.html** and is available as a portable document format (pdf) file. The publication is in the public domain and may be reproduced or reprinted in its entirety and without changes. A franking/imprint space for organizations and businesses is available on the lower half of the back cover. This publication was prepared by EPA's Indoor Environments Division (IED), Office of Radiation and Indoor Air (6609-J), 1200 Pennsylvania Avenue, N.W., Washington, D.C. 20460.

Single copies are available *free* from the following sources (multiple copies may be available in some instances; ask for details):

✓ State radon offices; see **http://www.epa.gov/radon/whereyoulive.html**.

✓ National Service Center for Environmental Publications (NSCEP) at 1-800-490-9198, **http://www.epa.gov/nscep/** or via email at nscep@bps-lmit.com.

Single or multiple copies are available for a *fee* from the See page 36 following sources (ask for details):

✓ The Conference of Radiation Control Program Directors (CRCPD) at (502) 227-4543 (multiple copy orders only).

✓ The American Association of Radon Scientists and Technologists (AARST) at (866) 772-2778 (multiple copy orders only).

✓ The National Radon Safety Board (NRSB) at (866) 329-3474 (multiple copy orders only).

EPA 402/K-13/002 | September 2013 (revised)

☐ *A Citizen's Guide to Radon: The Guide to Protecting Yourself and Your Family From Radon* (EPA 402/K-09/001, January 2009).
Provides basic information on radon, sources of radon, radon health risks, and how to test when you're *not* in a real estate transaction.

☐ *A Radon Guide For Tenants* (EPA 402-K-98-004, August 1998).
Provides tenants with basic information about radon, testing, and fixing. It also contains information directed to building owners and landlords. This document is only available online – **http://www.epa.gov/radon/pubs/tenants.html.**

Reducing Radon Levels In a Home

☐ *Consumer's Guide to Radon Reduction* (EPA 402-K-06-094, December 2006).
The consumer's basic source of information on how to reduce radon levels in a home's indoor air. It includes information about the key mitigation system components, installation and operating costs, radon health risks, and testing (when not in a real estate transaction).

Building a New Home to Be Radon-Resistant

☐ *Appendix F: Radon Control Methods (IRC, 2003).*
Published in the International Residential Code by the International Code Council (ICC) as a guide to building radon-resistant homes. Available from the ICC, 5203 Leesburg Pike, Suite 600, Falls Church, VA 22041-3401. Contact information: 1-888-ICC-SAFE, or via the Internet at http://www.iccsafe.org

☐ *Radon Control Methods (Section 49.2.5)*
Published in the National Fire Protection Association's (NFPA, 2003) Building Construction and Safety Code: NFPA 5000. NFPA, 1 Batterymarch Park, Quincy, Massachusetts 02169-7471. Contact information: 617 -770-3000, or via the Internet at www.nfpa.org

☐ *Standard Practice for Radon Control Options for the Design and Construction of New-Low Rise Residential Buildings* (ASTM E 1465-08, EPA 402-K-08-004*).
This consensus standard provides technical details on how to make radon-resistant features an integral part of a new home during construction. A must for builders or anyone building a new or custom home.

EPA 402/K-13/002 I September 2013 (revised)

Radon Technical Guidance

☐ ***Standard Practice for Installing Radon Mitigation Systems in Existing Low-Rise Residential Buildings*** (ASTM E 2121-03, EPA 402-K-03-007*). This consensus standard provides technical details on mitigating existing buildings. A must for professional mitigators.

☐ ***Protocols for Radon and Radon Decay Product Measurements in Homes*** (EPA 402-R-92-003, June 1993). This document is intended for use by qualified radon measurement technicians and testers, and laboratories that analyze radon devices and prepare radon test results reports. These protocols were written to guide routine radon measurements (*Citizen's Guide*) and those made in conjunction with real estate transactions (*Home Buyer's and Seller's Guide*).

☐ ***Indoor Radon and Radon Decay Product Measurement Device Protocols*** (EPA 402-R-92-004, July 1992). This document is intended for use by qualified radon measurement technicians and testers. It contains detailed technical information on the types of radon measurement devices, their proper use and maintenance, and quality assurance procedures. These protocols were written to guide routine radon measurements (*Citizen's Guide*) and those made in conjunction with real estate transactions (*Home Buyer's and Seller's Guide*).

*A single copy of ASTM E 2121 and E 1465 is free on request from EPA's National Service Center for Environmental Publications (NSCEP); 1-800-490-9198, http://www.epa.gov/nscep/, or via email at nscep@bps-lmit.com.

EPA 402/K-13/002 | September 2013 (revised)

9. State Radon Offices
(www.epa.gov/iaq/whereyoulive.html)

Up-to-date information on how to contact your state radon office is available on the web (above). You will also find a list of state hotlines, state indoor air coordinators, and state web sites (if available). Some states can also provide you with a list of qualified radon services providers. Native Americans living on Tribal Lands should contact their Tribal Health Department or Housing Authority for assistance.

EPA 402/K-13/002 | September 2013 (revised)

10. EPA Regional Offices

REGION	STATES	PHONE / FAX
US EPA New England/ **Region 1** One Congress Street, Suite 1100 John F. Kennedy Federal Bldg. Boston, MA 02114-2023	CT, MA, ME, NH, RI, VT	617-918-1630 617-918-4940-fax
US EPA/ **Region 2** 290 Broadway, 28th Floor New York, NY 10007-1866	NJ, NY, PR, VI	212-637-3745 212-637-4942-fax
US EPA/ **Region 3** 1650 Arch Street Philadelphia, PA 19103	DC, DE, MD, PA, VA , WV	800-438-2474 Toll-free 215-814-2086 215-814-2101-fax
US EPA/ **Region 4** 61 Forsyth Street, SW Atlanta, GA 30303-3104	AL, FL, GA, KY, MS, NC, SC, TN	404-562-9145 404-562-9095-fax
US EPA/ **Region 5** 77 West Jackson Blvd., (AE-17J) Chicago, IL 60604	IL, IN, MI, MN, OH, WI	312-353-6686 312-886-0617-fax
US EPA/ **Region 6** 1445 Ross Avenue (6PD-T) Dallas, TX 75202-2733	AR, LA, NM, OK, TX	800-887-6063 Toll-free 214-665-7550 214-665-6762-fax
US EPA/ **Region 7** 901 North 5th Street (ARTD / RALI) Kansas City, KS 66101	IA, KS, MO, NE	913-551-7260 913-551-7065-fax
US EPA/ **Region 8** 999 18th Street, Suite 500 (8P-AR) Denver, CO 80202-2466	CO, MT, ND, SD, UT, WY	800-227-8917 Toll-free 303-312-6031 303-312-6044-fax
US EPA/ **Region 9** 75 Hawthorne Street (Air-6) San Francisco, CA 94105	AZ, CA, HI, NV, GUAM	415-744-1046 415-744-1073-fax
US EPA/ **Region 10** 1200 Sixth Avenue (OAQ-107) Seattle, WA 98101	AK, ID, OR, WA	206-553-7299 206-553-0110-fax

11. Index

EPA 402/K-13/002 I September 2013 (revised)

EPA 402/K-13/002 I September 2013 (revised)

U.S. SURGEON GENERAL HEALTH ADVISORY

"Indoor radon is the second-leading cause of lung cancer in the United States and breathing it over prolonged periods can present a significant health risk to families all over the country. It's important to know that this threat is completely preventable. Radon can be detected with a simple test and fixed through well-established venting techniques." January 2005

Consumers need to know about the health of a house they are considering purchasing, including whether there is a radon problem, and if so, how to fix it. The *Home Buyer's and Seller's Guide to Radon* provides practical consumer information that every home buyer needs to know.

Consumer Federation of America Foundation

American Society of Home Inspectors

ENVIRONMENTAL
LAW·INSTITUTE®

Indoor Air Quality (IAQ)

www.ingramcontent.com/pod-product-compliance
Lightning Source LLC
Chambersburg PA
CBHW081305170526
45165CB00011B/3425